MATHEMATICS

REPORT OF THE
PROJECT 2061 PHASE I
MATHEMATICS PANEL

by David Blackwell
and Leon Henkin

AMERICAN ASSOCIATION
FOR THE ADVANCEMENT OF SCIENCE

1989

Founded in 1848, the American Association for the Advancement of Science is the world's leading general scientific society, with more than 132,000 individual members and nearly 300 affiliated scientific and engineering societies and academies of science. The AAAS engages in a variety of activities to advance science and human progress. To help meet these goals, the AAAS has a diversified agenda of programs bearing on science and technology policy; the responsibilities and human rights of scientists; intergovernmental relations in science; the public's understanding of science; science education; international cooperation in science and engineering; and opportunities in science and engineering for women, minorities, and people with disabilities. The AAAS also publishes *Science*, a weekly journal for professionals, and *Science Books & Films*, a review magazine for schools and libraries.

ISBN 0-87168-344-X

AAAS Publication 89-03S

Library of Congress Catalog Card Number: 89-103

© 1989 by the American Association for the Advancement of Science, Inc., 1333 H Street NW, Washington, D.C. 20005

All rights reserved. No part of this book may be reproduced or transmitted in any form or by any means, electronic or mechanical, including photocopying or recording, or by any information storage and retrieval system, without permission in writing from the Publisher.

Printed in the United States of America

CONTENTS

	Page
ACKNOWLEDGMENTS	iii
PHASE I MATHEMATICS PANEL	v
FOREWORD by F. James Rutherford, Project Director, Project 2061	vii
PREFACE by David Blackwell and Leon Henkin, Cochairs, Mathematics Panel	xi
SECTION 1: INTRODUCTION	1
SECTION 2: THE PROCESSES OF MATHEMATICS	5
Abstraction/Representation	5
Symbolic Transformation	6
Application/Comparison	11
SECTION 3: THE SUBJECT AREAS OF MATHEMATICS	15
Arithmetic	16
Algebra	18
Geometry	20
Analysis	22
Discrete Mathematics	23
Logic and Set Theory	25
Probability and Statistics	27
A Sample Problem	28
SECTION 4: MATHEMATICS, SCIENCE, AND TECHNOLOGY	31
SECTION 5: MATHEMATICS AND LANGUAGE	33
SECTION 6: EMOTIONS AND MATHEMATICS	37
SECTION 7: CONCLUDING REMARKS	41
POSTSCRIPT	43
APPENDIX: MATHEMATICS PANEL CONSULTANTS	47

ACKNOWLEDGMENTS

On behalf of the Board of Directors of the American Association for the Advancement of Science, I wish to acknowledge with gratitude the many useful contributions made by the members of the Phase I Mathematics Panel to the first stage of Project 2061.

The eight panel members were most generous with their time and efforts over a two-year period in developing their response—as presented in this report—to the complex question of what young people should know about mathematics by the time they complete their high school education. The board is also very grateful to David Blackwell and Leon Henkin, who cochaired the panel and wrote the panel report.

During this essential first stage of Project 2061, the Mathematics Panel was one of five scientific panels charged by the AAAS with developing independent reports on five basic subject-matter areas. At the same time, the Project 2061 staff—in conjunction with the National Council on Science and Technology Education—was preparing a separate overview report—*Science for All Americans*—that was able to draw on the conclusions reached by the individual panels.

We also want to extend our thanks to the many people who assisted the Mathematics Panel in the course of its deliberations—the consultants, the national council members and the other reviewers, and the Project 2061 staff.

Finally, it is appropriate to note that Project 2061 is indebted to the Carnegie Corporation of New York and the Andrew W. Mellon Foundation for their overall and ongoing support of our various Phase I efforts.

Sheila E. Widnall
Chair, Board of Directors, American Association for the
   Advancement of Science

PHASE I MATHEMATICS PANEL

David Blackwell (Panel Cochair) Professor of Statistics and Professor of Mathematics, University of California, Berkeley

Leon Henkin (Panel Cochair) Professor of Mathematics, University of California, Berkeley

Lenore Blum Professor of Mathematics, Mills College; Research Scientist, International Computer Science Institute

Paul Garabedian (Physical and Information Sciences and Engineering Panel Liaison Member) Professor of Mathematics, Courant Institute of Mathematical Sciences, New York University

Paul Halmos Professor of Mathematics, University of Santa Clara

Harvey Keynes (Technology Panel Liaison Member) Professor of Mathematics, University of Minnesota at Minneapolis

R. Duncan Luce (Social and Behavioral Sciences Panel Liaison Member) Distinguished Professor of Cognitive Science and Director of the Irvine Research Unit in Mathematical Behavioral Science, University of California, Irvine

Ingram Olkin Professor of Statistics and Professor of Education, Stanford University

James Sethian Associate Professor of Mathematics, University of California, Berkeley

Audrey Terras (Biological and Health Sciences Panel Liaison Member) Professor of Mathematics, University of California, San Diego

P. Emery Thomas Professor of Mathematics, University of California, Berkeley

Panel Production Staff: Sara Wong, University of California, Berkeley

FOREWORD

This report is one of five prepared by scientific panels as part of Phase I of Project 2061. Each of the panel reports stands alone as an independent statement of learning goals in a particular domain. In addition, the reports contributed to *Science for All Americans*, a Phase I report that cuts across all of science, mathematics, and technology.

The work of the Mathematics Panel was to reflect on all aspects of mathematics—its nature, principles, history, future directions, social dimensions, and relation to science and technology—and to produce a set of recommendations on what knowledge and skills are needed for mathematical literacy (as part of general scientific literacy). The other panels focused in a similar way on the biological and health sciences, the physical and information sciences and engineering, the social and behavioral sciences, and technology.

In considering this report, it is helpful to see it in the context of Project 2061 and to be aware of the manner in which it was generated.

The American Association for the Advancement of Science initiated Project 2061 in 1985, a year when Comet Halley happened to be in the earth's vicinity. That coincidence prompted the project's name, for it was realized that the children who would live to see the return of the comet in 2061 would soon be starting their school years. The project was motivated by concern that many share for the inadequate education those young Americans will receive unless there are major reforms in science, mathematics, and technology education.

Scientific literacy—which embraces science, mathematics, and technology—has emerged as a central goal of education. Yet the fact is that general scientific literacy eludes us in the United States. A cascade of recent studies has made it abundantly clear that by national standards and world norms, U.S. education is failing too many students—and hence the nation. The nation has yet to act decisively enough in preparing young people—especially the minority children on whom the nation's future is coming to depend—for a world that continues to change radically in response to the rapid growth of scientific knowledge and technological power.

Believing that America has not more urgent priority than the reform of education in science, mathematics, and technology, the AAAS has committed itself, through Project 2061 and other activities, to helping the nation achieve significant and lasting educational change. Because the work of Project 2061 is expected to last a decade or longer, it has been organized into three phases.

Phase I of the project has established a conceptual base for reform by defining the knowledge, skills, and attitudes all students should acquire as a consequence of their total school experience from kindergarten through high school. That conceptual base consists of recommendations presented in *Science for All Americans* and the five panel reports.

In Phase II of Project 2061, now under way, teams of educators and scientists are transforming these reports into blueprints for action. The main purpose of the second phase of the project is to produce a variety of alternative curriculum models that school districts and states can use as they undertake to reform the teaching of science, mathematics, and technology. Phase II will also specify the characteristics of reforms needed in other areas to make it possible for the new curricula to work: teacher education, testing policies and practices, new materials and modern technologies, the organization of schooling, state and local policies, and research.

In Phase III, the project will collaborate with scientific societies, educational associations and institutions, and other groups involved in the reform of science, mathematics, and technology education, in a nationwide effort to turn the Phase II blueprints into educational practice.

Each of the five panels was composed of 8 to 10 scientists, mathematicians, engineers, physicians, and others known to be accomplished in their fields and disciplines and to be fully conversant with the role of science, mathematics, and technology in the lives of people. The panelists were different from one another in many respects, including their areas of specialization, institutional affiliations, views of science and education, and personal characteristics. What made it possible to capitalize on the rich diversity among the panelists was what they had in common—open minds and a willingness to explore deeply the questions put to them.

The basic question put to the Mathematics Panel was: What is the mathematics component of scientific literacy? Answering this question—difficult enough in itself—was made more difficult by the conditions, or ground rules, set by Project 2061. Abbreviated here, these were:

- *Focus on mathematical significance.* Identify only those concepts and skills that are of surpassing mathematical importance—those that can serve as a foundation for a lifetime of individual growth.

- *Apply considerations of human significance.* Of the knowledge and skills that meet the criterion of mathematical significance, select those that are most likely to prepare students to live interesting and responsible lives. Individual growth and satisfaction need to be considered, as well as the needs of a democratic society in a competitive world.

- *Begin with a clean slate.* Justify all recommendations without regard to the content of today's curricula, textbooks, state and school district requirements, achievement tests, or college entrance examinations.

- *Ignore the limitations of present-day education.* Assume that it will be possible to do whatever it may take—design new curricula and learning materials, prepare teachers, reorganize the schools, set policies, or locate resources—to achieve desired learning outcomes.

- *Identify only a small core of essential knowledge and skills.* Do not call on the schools to cover more and more material, but instead recommend a set of learning goals that will allow them to concentrate on teaching less and on doing it better.

- *Keep in mind the target population—all students.* Propose a common core of learning in mathematics that can serve as part of the educational foundation of all students, regardless of sex, race, academic talent, or life goals.

Taking these ground rules into account, the members of the Mathematics Panel met frequently over a period of nearly two years to present and debate ideas and to consider the suggestions of consultants. The panel members prepared working papers and revised them in response to the criticisms of reviewers. This process—which also included meetings with the chairs of other panels—led to the preparation of this report.

The task ahead for the United States is to build a new system of education that will ensure that all of our young people become literate in science, mathematics, and technology. The job will not be achieved easily or quickly, and no report or set of reports can alter that. I believe, however, that this report on mathematics literacy, along with the other panel reports and *Science for All Americans*, can help clarify the goals of elementary and secondary education and in that way contribute significantly to the reform movement.

F. James Rutherford
Project Director, Project 2061

PREFACE

One of the ground rules mentioned by Project Director F. James Rutherford in his foreword is "Keep in mind the target population—all students." We took that admonition seriously, and we hope that you will keep it in mind while reading our report. (See our postscript for a fuller discussion of this point.)

Another precept that we took from the charge given to the Project 2061 panels was "Don't come up with a laundry list of topics to be covered." This made us hold down the number of topics mentioned in Section 3 of our report, "The Subject Areas of Mathematics," and we hope that no one will interpret our recommendations as indicating that the topics we do mention are exactly the ones that must be covered in school curricula. To the contrary, the essential ideas of mathematics outlined in Section 2, "The Processes of Mathematics," must and can be exemplified in a wide variety of topics, from which individual teachers should select to fit the needs and interests of their students.

This report, written by the undersigned, was generated in a series of intense discussions about the essential ideas of mathematics that everyone should know and understand—discussions carried on within daylong meetings of the Mathematics Panel during 1985 and 1986. Project 2061 staff members Chick Ahlgren, Jim Rutherford, and Pat Warren joined us on several of those occasions, as did our invited consultants, and all of these individuals helped to fuel the vigorous disagreements, forcefully expressed, out of which some semblance of consensus would finally appear. Still, in such matters there is no final word, and we authors have to assume responsibility for what we have set down.

Drafts of this report were discussed at several meetings of Project 2061 personnel and at national meetings of the American Association for the Advancement of Science and of the Mathematical Association of America, and they were also reviewed by individual mathematicians. Although some changes based on these discussions and reviews have been incorporated in the present text, the contents of the report remain our responsibility, in that we are the ones who wrote the report. We hope that this report will spark continuing debate among mathematicians, educators, teachers, and the public—debate that we hope will contribute to the quality and liveliness of the teaching and learning of mathematics.

David Blackwell and Leon Henkin
Cochairs, Mathematics Panel

SECTION 1
INTRODUCTION

This report responds to a single question: "What are the important ideas of mathematics that people should know and understand by the age of 18?" To communicate our response, we must first discuss the meaning of the word "mathematics." We do not intend a mathematical definition of the word "mathematics" such as one finds for technical words, like "polygon" in geometry or "polynomial" in algebra. Rather, we seek to communicate a sense of what mathematics *is* to mathematicians—a sense much broader than that which most readers of our report will have gained from their own experience with mathematics.

It is essential that you, the readers, and we, the authors, of this report realize that this difference in experience *with* mathematics will lead most readers to expect a very different kind of answer than we shall present to the starting question that has brought us together. This broader experience of mathematics that we have acquired through our profession is *not* something very remote and complex that can be understood only after years of abstruse study. On the contrary, it informs the simplest mathematical subjects to the same extent as the more sophisticated ones. For this reason, the "important mathematical ideas" that we present below have the potential to revolutionize the nature of teaching and learning mathematics through all the years during which students grow to be 18 years of age. But this can occur only if the public comes to realize what a limited part of mathematics is now experienced by most people.

We have been addressing readers of this report, but of course they are not homogeneous in their mathematical experience. Among groups having distinctive perspectives regarding mathematics are "typical adults," who have reached but have not studied mathematics beyond high school; teachers of mathematics at all school levels; and users of college-level mathematics in business, technology, or pure or applied science (whether physical, biological, or social). Many readers will not fit into any of these groups, but we hope that all of them will find parts of our report that are of special interest to them.

What does the word "mathematics" mean for those we've called typical adults (TAs, for short)? Since

TA = person over 18 with no math beyond high school,

we must examine computation, which is the principal ingredient of school mathematics courses. Computation in school begins with whole numbers, then deals with fractions and negative numbers, and later involves algebraic and possibly trigonometric formulas. But always the rules are given by teachers and textbooks, have to be "learned"—that is, memorized—and form the basis of "skills" that consist of the rapid application of the rules to given numbers or formulas, with few errors. Although some explanation of the rules is given, students can and do succeed in standard tests without understanding them. And it is rare for students to experience how the rules grew out of ordinary experience, or to be told how they were found in stages, through history.

There are certainly components of school mathematics other than the ones described above, but memorized computation looms so large in the total classroom experience that it forms the core of what TAs consider mathematics to be. For mathematicians, however, memorized computation is a necessary but very small part of the total mathematical experience. For this reason, we feel that the TA's view is much too narrow to provide an adequate framework for "the important ideas of mathematics" that we seek to identify.

Actually, there are various areas of experience *outside* the classroom in which TAs develop what we would call mathematical ideas—even though the TAs themselves do not classify them as mathematics.

Let us consider an example. Suppose that two youngsters set out to share equally a pile of money consisting of four dollar bills and a bunch of coins worth a dollar and a half. If they are suitably experienced, they may find it easy to agree on taking $2.75 each. A TA observing them may see that they have solved a problem related to division of 5½ by 2, and conclude that the youngsters have applied mathematics as taught in school.

In reality, the youngsters will quite likely never have sought to apply a memorized division algorithm. It is more likely that they have called on basic rules of exchange they have developed from their own money-handling experience (for example, a quarter is worth two dimes and a nickel). They will have used principles that they abstracted from long practice in sharing (which a mathematician might describe in terms of technical concepts, such as "injective mappings," that are *certainly* not in the youngsters' minds). And they will have combined these principles by means of algorithms involving the exchange of coins that they have themselves developed but cannot formulate verbally.

Our view is that these processes of defining rules of exchange, abstracting principles of sharing, and developing algorithms, are *essential* elements of "doing mathematics."

So the TA and we mathematicians, watching the youngsters sharing some money, agree that mathematics is being used. But the TA and we mean quite different things as we nod agreement to one another.

Through this example we have attempted to indicate—admittedly somewhat obscurely—some of the components that must be joined to rote computation to obtain what we call *mathematics*. We call these components *mathematical processes*. It is appropriate now for us to give a fuller list of these processes, with brief comments about each, and then discuss how they interact among themselves within ongoing mathematical experience (Section 2 below). This description of processes itself constitutes one part of the "important ideas of mathematics" that we believe all 18-year olds should come to understand. In a subsequent part of the report (Section 3) we indicate by example how these interactive mathematical processes can be seen at work in a variety of subject areas of mathematics. In Section 4, we show how they emerge from and reflect back upon realms of *nonmathematical* experience—including, but not limited to, those that form the subject of the various areas of science and technology. In Section 5, we comment on the special ways in which language is used in mathematics and the ways in which mathematics functions as the language of science and technology. In Section 6, we comment on the relevant emotional contexts within which mathematics is experienced.

Each of these six sections of our report contributes to our total answer to the starting question: "What are the important ideas of mathematics that people should know and understand by the age of 18?"

Before ending these introductory remarks, we pause to make two points that may help orient the readers as they move through Sections 2 through 6 below.

Henceforth, we continue to use "TA" to refer to a "typical adult"—one who has passed the age of 18 and has not studied mathematics beyond high school—but there is a subtle shift in our meaning. Whereas we write above about TAs who have come through school mathematics as it is now, we write below about the TAs of the future, imagining that they have had the benefits of a schooling that provides a full range of mathematical activities without the present undue emphasis on computing according to memorized rules. If readers feel at times that our expectations are above the normal ability range for TAs, they should be assured that we have not fashioned our prescriptions for mathematical specialists or even for students who are college-bound. Rather we have sought to describe mathematical ideas that we believe are well within reach of 95 percent of the population—*provided they are furnished with a proper school environment.* (We return to this theme in a postscript at the end of this report.)

On another issue, we are aware that some readers will recall that in an earlier decade, the school mathematics curriculum was subjected to substantial changes—originating with mathematicians—that became known popularly as the new math. There is a residue of resentment and suspicion left over from that experience, and we wish to make it clear that our present report is intended to lead in a very different direction from new math. The essence of the earlier movement was an attempt to explain to students the computing rules that they had to memorize. This was done by using a means of explanation that appealed to mathematicians—deducing the rules from certain general laws about numbers (see below). The new math's emphasis on deduction, without a concomitant treatment of the mathematical processes of abstraction and application, undercut the possibility of relating the rules of computing to children's nonmathematical experience. The use of deduction to explain computational rules to students also overlooked an important principle of learning—namely, that for any learner it is essential to explain new ideas in terms that are already familiar from the learner's previous experience. However, in this report, authored by mathematicians in Phase I of Project 2061, ideas are presented that will be transformed into curricular changes in Phase II by people whose professional work makes them very familiar with the kinds of explanations that relate to the early experience of children.

SECTION 2

THE PROCESSES OF MATHEMATICS

Mathematics, as well as science and technology, have developed in the course of history from ordinary human experience. In the life of each individual there are many natural opportunities for such development to occur. Instruction in mathematics must seize upon, nurture, and advance this kind of development. This process can help youngsters realize their potential for experiencing and understanding what mathematics is, and how it can help them achieve goals in realms of their nonmathematical activities.

We see a cycle of fundamental mathematical processes that recur over and over, consisting in its most basic form of abstraction, symbolic transformation, and application. This cycle not only occurs at the interface of ordinary and mathematical experience but is repeated many times *within* the realm of mathematics—leading the subject to greater levels of generality and hence on to greater possibilities of powerful effect.

ABSTRACTION/REPRESENTATION

The process of abstraction begins with the noticing of similarity between two or more objects, events, or situations. It then proceeds to recognition of some feature of this similarity that can be looked for in other objects (or events or situations). After that, it selects some word, symbol, picture, or other object to "represent" the common feature of the similar objects.

This process of abstraction is well in place by the time children reach the age of 2, when they are successfully using words—such as "dog" or "blue"—to link particular objects with others sharing a similar aspect. The word "round" is another abstraction from observed objects, but the word "circle"—as it finally comes to be used in a geometry course—involves several abstractions following one another, of which "round" is only the first.

The whole numbers are first abstracted from batches of objects through operations of counting that involve fingers, eyes, and voice in patterns of imitated ritual. Later on, the operation of addition is abstracted from operations of manipulation such as combining two batches of objects to form one big batch, laying two sticks end to end along a ruler, or timing two consecutive events successively. Much later, the abstract operation of addition is itself a particular object that enters with other such objects into a further abstraction—namely, the general mathematical notion of an operation.

We have given examples of abstractions, and we must now say a word about the *representations* that are produced in connection with them—most commonly, words or other symbols. Such symbols are of great use to an individual in that they enable the person to proceed with reflective *thoughts* about the abstract concepts, relating those concepts to various concrete objects as well as to other abstractions. However, the association of symbols with abstractions is absolutely *essential* for communication with others about the abstractions. There is no way to point with a finger at an abstraction; but one can point at a concrete symbol that represents it. If the symbol has a *shape* such as a bar graph, it may greatly facilitate the symbol's use in communicating ideas about the abstraction represented. As individuals gain practice in using mathematical language, they learn that strings of symbols—such as formulas, equations, and sentences—have a shape too. Even a particular symbol, such as our expression "TA," may have a shape that helps us to think about its meaning—"typical adult."

It is important to keep in mind that although each symbolic representation of an abstraction stands for an aspect that several objects or situations have in *common*, it ignores the *differences* between those objects or situations. This allows us to concentrate attention on certain features of experience, by relieving us of the need to keep other features continually in mind. The resulting economy of effort is very useful—providing that in our abstraction, we have not ignored features that play a significant role in determining the outcome of the events we are studying. Because of uncertainty as to the adequacy of our abstractions, mathematical work must always be brought back to the realm of experience from which it emerged, to judge its adequacy.

SYMBOLIC TRANSFORMATION

After abstractions have been made and symbolic representations of them have been selected, the symbols become objects that enter into a variety of mathematical activities. They can be combined and recombined in various ways, both by themselves and in combinations with words and other symbols. Sometimes this is done with a fixed goal in mind; at other times, it may be done in the context of play or experiment "to see what will happen."

Symbols and words are formed into meaningful "strings" of two kinds. One kind consists of strings that represent objects and are called *terms*. Examples are "$(7 + 5)$", "the largest two-digit integer", and "the shape of the shortest closed curve in a plane, P, that surrounds a region of P having an area of 5 square centimeters". Sometimes, the object denoted by a term can be identified easily from the meaning of the constituent words and symbols; in other cases, the denotation of a term has to be worked out by a process of computation—a process that may or may not be known in advance, and that may take anywhere from seconds to years to complete.

The other kind of string consists of strings of words and symbols that express propositions that are either true or false. These are called *sentences*. Examples are "$(7 + 8) > (36 - 18)$" and "For every even number x greater than 2 there are prime numbers p and q such that $p + q = x$". As with terms, so with sentences: Sometimes the truth value (truth or falsity) of a given sentence can be identified easily from the meanings of the constituent words and symbols; at other times, it has to be worked out by a process of deduction using methods that may be already known or may have to be invented.

What motivates us to work at obtaining the value of a particular term or the truth value of a particular sentence? If it is not a question posed by someone else (the usual situation in a classroom), there are several ways in which we may launch such a project ourselves. In a simple case, the question may lead directly to a solution of some problem in a domain of nonmathematical experience that led us to undertake the mathematical activity. In a more complex case, we may be groping toward the solution of such a problem, and it may occur to us that if we could determine the value of some particular term or sentence, this might open the way to arrive at our solution by some further sequence of mathematical steps. This kind of testing of hunches may occur many times in a tangle of efforts to find our way toward the solution of a difficult problem. Even when we succeed, the number of incorrect hunches along the way may exceed the number of correct ones.

Another kind of context leading us to seek the truth value of some sentence occurs if we have found that several objects of a certain kind have some property, and we wonder whether another object resembling them also has that property. Eventually, we may suspect that *all* objects of that kind have the particular property and try to settle the question. If we succeed and find a positive answer, we will have carried through a process of *induction*.

A related mathematical process is that of *generalization*. Typically, we may find that a certain form of mathematical law holds for each of several systems, and we may seek to determine a whole class of systems in all of which this law will hold. A successful outcome to such a request may lead to a new abstraction, if we introduce a new word or symbol to represent the class of all the systems for which we are able to show that the law holds.

The processes of computation, deduction, testing hunches, attempting inductions, and seeking generalizations, are each undertaken with some idea in mind as to the nature of the finding that may ensue. However, some valuable activity in mathematics—just as in the other sciences and in the arts—may begin without any goal in mind other than to rearrange things in new ways, to see whether anything results that might interest us.

All of the mathematical processes mentioned above involve manipulation of the symbols representing our abstractions, and they lead us to findings that we can express by means of formulas and sentences that employ those symbols. We have thus grouped these processes under the heading of Symbolic Transformation. Among these, we now single out two for extended description: *deduction* and *computation*. These two processes take up much more time, within mathematical activity, than any of the others. Furthermore, each has been made the *subject* of mathematical investigation in a part of mathematics (called *metamathematics*) in which mathematical methods have been used to investigate mathematics itself.

Deduction

When words and symbols representing various abstract and concrete objects are formed into a suitable string, they constitute a sentence. This expresses some proposition about the objects referred to and may be true or false. Often, in mathematics, we do not *know* whether a particular sentence is true or false, and so we make efforts to find out. This is where the process of deduction comes in.

For sentences that are entirely about concrete objects that we can see, hear, or touch, we can in principle decide on truth or falsity with our eyes, ears, or fingers. But for sentences that are partly or wholly about abstractions, we must bring in other tools. In this process, the logical rules of definition and deduction play an essential role.

The usual situation is that while we are wondering about the truth value of a particular sentence, S, we *know* that some other sentences—say, P and Q—*are* true. If we pass from P and Q through a sequence of intermediate sentences each obtained from earlier ones by laws of logic, and arrive at S, then we can be sure that S too must be true. Such a sequence of steps is called a *deduction* of S from P and Q. P and Q are the *premises* of the deduction; S is the *conclusion*.

Although the laws of logic are related to truth in the way described, they are formulated only in terms of the shapes, or forms, of the sentences involved. For example, there is a law (from the logic of equality) that allows us to pass from the premise $x + 6 = 9$ to the conclusion $x + 6 - 6 = 9 - 6$, *even though we do not know* whether the symbol "x" represents the number 3 (which would make the premise true) or some other number (in which case the premise would be false).

Hence we may be able to obtain a correct deduction of a sentence, S, from premises P and Q, even when we're unsure of the truth of P or Q or both. This procedure is often very useful in mathematics. In the case where we *know* that premises P and Q are true, a deduction of S from them is called a *proof* of S.

We now come to a very common misunderstanding about the deductive element of mathematics. It deals with our efforts to *find* a deduction leading from some premises—say, P and Q—to a conclusion, S. Many people believe that the successive applications of laws of logic that give us the steps of such a deduction are supposed to guide their mental states, or thoughts, in a step-by-step fashion, starting with P and Q and ending with S. They believe that writing down the deduction will simply be a recording of their successive mental states in this process. If they have difficulty in ordering their thoughts so as to pass from one to another by a law of logic, they fear that they don't have "a logical mind." *This idea, however, is wrong.*

A proof is a symbolic device that—once found—enables us to check the truth of the conclusion; it is *not* a report of mental processes by which we *discover* that truth. The laws of logic relate the form of a conclusion-sentence to the form of premise-sentences; they do *not* describe thought processes. There are rules to check whether a proposed proof is correct, but *there are no general rules for finding proofs.*

When a mathematician sets out to find a proof to solve an interesting problem, all sorts of strange ideas may follow one another in his or her mind. Many of these ideas are likely to be intuitive guesses that are wrong; some of them may turn out to be steps in the proof finally found, but they may come to mind out of order. It is just because there are no rules for finding proofs, in general, that mathematicians prize *some* proofs beyond all wealth.

Of course, at each grade level in school, there are some very limited kinds of problems for which the necessary deductions are so simple that it would be possible to give rules for finding them. Children sometimes ask for such rules so that they can be relieved of the difficulty of searching and struggling to reach conclusions that lie beyond rules previously given. But such struggling is an essential part of mathematical experience; students must learn to solve problems and make deductions that lie beyond the rules they know, at increasing levels of difficulty.

To memorize lots of special rules for constructing deductions for special kinds of problems that are sprinkled through the curriculum places the main burden of learning on memory, undercuts general understanding, and leaves the student unprepared to cope with new kinds of problems that may be encountered outside of school—problems that may *not* be covered by the standard curriculum. Students must learn instead to develop intuitive pictures of how things work in the domain they're studying, and to seek among those intuitions to find the needed deductions. They must do lots of guessing to generate intuitive ideas, realizing that each guess must be tested for correctness and that wrong guesses can generally be used to produce better guesses.

We have been writing of individual deductions that are needed in solving problems, but there are also whole cascades of deductions that are used in creating a theory. Typical adults (TAs) at present have usually had a limited look at this phenomenon if they have studied an axiomatic theory of plane geometry, such as Euclid originated.

Whenever there is a large, cohesive mass of related mathematical findings, they can be *organized* by picking out a small number of them and then showing how all of the others can be obtained from these by successive deductions. The starting propositions are called *axioms* and the deduced ones are called *theorems.* There is never a *unique* set of the propositions that must function as axioms; different axiomatizations of the same set of facts are often useful for different purposes.

Just as the *sentences* of an axiomatic theory are organized by chains of deductions, so the *terms* that enter into these sentences, which express the concepts that the propositions are about, are organized by chains of mathematical definitions. The starting terms are not defined within the theory (just as the axioms are not proved); they are called the *primitive terms* of the theory. As in the case of axioms, there are different ways to choose primitive terms in setting up a theory. When terms are left undefined in constructing a given theory, it allows them to be defined in different ways when different applications of the theory are made. Giving definitions to the primitive terms of an axiomatic theory is called *interpreting* the theory. Many interpretations are possible for the same theory.

In the traditional high school geometry course, an axiomatic theory is presented, students are asked to follow the deductions of some theorems from the axioms (that is, the proofs of the theorems), and they are then asked to construct some proofs of their own. It is desirable, however, for students also to gain experience in constructing their own axiom systems in contexts in which they have already gained familiarity with a large number of the basic facts. This activity will enable them to see the power and economy that can be obtained by organizing these facts, as well as to appreciate the elegance of reducing a complex set of facts to a few. (A simple example would be the problem of developing an axiom system to deductively organize all of the laws of inequalities $<$, $\leq$, $>$, and $\geq$ that hold in the system of integers.)

When the axioms of a theory are familiar, or at least seem to be true, the student gains security in seeing other propositions deduced from the axioms. One of the disconcerting elements of the new-math program was the deduction of the correctness of traditional computational algorithms from axioms for the system of natural numbers. One difficulty was that some of these axioms embodied ideas that were neither familiar nor evident to the students.

Computation

Is the sentence $2 + 3 = 9 - 3$ true or false? To answer this, we compute the values of the expressions $2 + 3$ and $9 - 3$. These expressions are examples of mathematical formulas of a kind called *terms*. Almost every mathematical sentence has one or more component parts that are terms. Thus, computation and deduction are closely interwoven processes in mathematics. Yet somehow in the school curriculum, they are often kept far apart.

There is one aspect of computation with which every TA has had years of experience: applying prescribed rules to transform one expression—say, $27 + 34$—to obtain another expression—in this case, 61. In general, these rules enable us to take any expression consisting of two numerals (representing whole numbers) separated by an operation symbol (such as $+$ or $\times$) and to transform it to obtain one numeral—which represents the number resulting when the operation (addition or multiplication) is performed on the numbers represented by the *given* numerals.

Any prescribed set of rules enabling us to carry out a fixed kind of computation on arbitrary terms of a given sort is called an *algorithm*. Usually, if the given terms represent numbers, then the outcome of the computation also represents a number (as above). However, this is not always the case; for instance, the outcome could represent a truth value (truth or falsity). For example, there is an algorithm that starts from any given pair of fractions, such as $4129/3077$ and $951/709$, and produces a truth value that is T if the fractions are equal (that is, they represent the same point on a number line), and F if they are unequal. A truth-valued algorithm like this one is often called a *decision-method*: In our example, the algorithm can be used to *decide* whether or not two given fractions are equal.

Many TAs will not have seen an algorithm described in their school days for deciding whether two fractions are equal; of those who will have seen it, many will have forgotten it. Nevertheless, with some effort, many TAs will be able to *devise* such an algorithm. In fact, there are several different algorithms for this decision-problem that they could devise, some being much easier to apply than others.

This is an important kind of mathematical activity—to *invent* an algorithm, instead of just being handed one and told to use it to "grind out numbers." The construction of algorithms is an essential part of mathematical experience; the use of someone else's algorithm to grind out numbers over and over is more and more assigned to electronic calculators. Finding a program for a computer is a form of inventing an algorithm, providing an important activity that links mathematics with computer science.

As with the general problem of finding deductions, so with the general problem of inventing algorithms for a given class of computations: There are no rules that lead in a guaranteed way to the desired solution. But at every level of learning there are problems of algorithm construction that are accessible, can lead to rich mathematical experience and deeper understanding, and can be used in group work to improve mathematical communication skills. Well-set problems of this kind will inevitably lead students to find connections between deduction and computation, and will lead them to *experiment*.

When algorithms are imposed by textbooks or teachers, students—and hence the TAs who evolve from them—are given the idea that to "know" mathematics is to be able always to proceed straight toward the right answer. But when mathematics arises in "a natural setting," it always leads to guessing, checking, discarding, simplifying and then complicating, seeking similarities—all of which are called into play in the search for computational algorithms.

The processes of deduction and computation described above, as well as the other processes of symbolic transformation mentioned earlier, are employed repeatedly in an ongoing sequence, during extended mathematical activity, once symbolic representations of abstractions have emerged from a domain of experience encountered earlier. Ultimately, however, we must bring the results of our symbolic transformations back to bear upon the original domain that led to our abstractions.

APPLICATION/COMPARISON

We now come to the final element in our cycle of mathematical processes. We began with some objects (or events or situations), made abstractions and represented them (in some symbolic manner), and used the representations to carry through an intertwined sequence of deductions, computations, and perhaps other symbolic transformations. We arrive in this way at new sentences involving the representing symbols, and so we return to the original objects that caught our attention to see what information about them we can derive from these new sentences.

In a "baby problem," such as pushing together a pile of three stones and a pile of four stones, we may get exactly the information we are seeking—the new big pile has seven stones. But sometimes at a more advanced level, the information is *not* exactly what we are seeking. For instance, in a physics class we may try to compute the time needed for a certain weight to slide down a smooth inclined surface, but our computed answer fails to match the time measured with a stopwatch: What is wrong? Maybe the computation was done with laws of motion that "neglected" frictional forces. It doesn't mean we forgot, or didn't know, about friction. It could mean that we knew the surface was "quite smooth" and thought that the slowing effect of the friction wouldn't affect our stopwatch measurement, so we saved computing time by using the simpler, frictionless laws of motion.

Or in a biology experiment, we may have placed a few cells in the center of a petri dish and tried to compute how large an area will be covered by the cell colony in 24 hours using the information that this kind of cell doubles every 80 minutes. We know that cell division slows down as the cells approach the edge of the petri dish, so the simple formula based only on the doubling time may give a colony size a bit larger than will really occur. But we're not sure *how* close to the edge the slowdown occurs, and we don't know just how to modify the simple formula when we do get close enough. So we go ahead and use the simple formula, aware that our computation gives only an approximate answer. And then we wonder whether the "error"— the difference between the computed value and the measured value—is likely to be less than 2 percent of the latter.

The process of setting up mathematical laws as a basis for predicting real-world events by deduction and computation is called "modeling." Such mathematical models are always susceptible to giving predictions that are literally false but may be close enough to the truth to be very useful. The question of how close is close enough often has an answer that goes beyond science or mathematics. It depends on the use to which the prediction is to be put, and on how important the user feels it is to succeed in that use. The general situation is that the more effort and resources one is willing to commit to modeling, deduction, and computation, the more accurate the prediction that can be made. Sometimes, we begin consideration of a problem by setting a limit to the inaccuracy we are willing to tolerate in the answer, and then we try to make an estimate of the cost and the likelihood of our being able to obtain mathematically an answer that meets these requirements. In some cases, this estimate may lead us to abandon the project.

So far, we've been writing about the information we get by applying the results of a mathematical investigation to the objects (or events or situations) that gave rise to the investigation. Because an abstraction links different objects (events or situations) that have something in common, it is often possible to use a *single* mathematical result, or theory, to obtain information about several different situations. This linking of diverse areas of experience by abstract mathematical statements that express their commonality often leads us to use what we already know about one area to discover new information about another quite different domain.

A particularly exciting mathematical event occurs, occasionally, when a model that has been abstracted from several related situations is unexpectedly found to be equally valid for another situation that was thought to be extremely different. For example, a geometric theory of convex shapes turned out to provide a way to find optimal strategies in the theory of games—which in turn is used to analyze economic behavior.

SECTION 3

THE SUBJECT AREAS OF MATHEMATICS

The cycle of mathematical processes described above characterizes important interacting phases of mathematical experience. Actually, each process—abstraction, symbolic transformation, and application—can be found in every empirical science and also, to at least a rudimentary extent, in the ordinary business of daily life. However, the processes are of the essence in mathematics because that is where the cycle is repeated over and over. The concepts abstracted in one cycle become the objects from which a further abstraction is made in the next cycle; the symbols used to represent abstractions at one level are combined in complex formulas that are used to define a single symbol representing an abstraction at the next level. Thus, the objects of a particular mathematical investigation may be many times removed from those that are familiar through our senses, and the mathematical language employed may contain terms that can be pursued only by following a long train of ascending definitions.

Nevertheless, each of the major subject areas of mathematics contains material sufficiently elementary to have come to the attention of our TAs—the typical adults who have not studied mathematics beyond high school.

The mathematical subject areas likely to be most familiar to TAs are *arithmetic*, *algebra*, and *geometry*. Trigonometry is commonly encountered as a high school course, but it is not nearly extensive enough to rate as a major subject area of mathematics; rather, it is only a patch on the border between algebra, geometry, and another subject area called *analysis*. The real starting point of analysis is calculus, now studied in high school by only a small fraction of students. Other mathematical subject areas of which only small bits currently occur in school courses are logic and set theory, probability and statistics, and discrete mathematics. Logic and set theory are often called "the foundations of mathematics" because it is possible to organize all the various subject areas of mathematics into one giant axiomatic theory whose primitive terms and axioms come only from logic and set theory. (Although of theoretical interest, however, this reduction is not especially significant for TAs.)

We now list, for each subject area of mathematics, what seem to us to be the concepts, theorems, methods, and skills (abilities) that ought to be known and understood by TAs whose life span will lie mainly in the period between now and the year 2061. However, before proceeding to lay out our lists, we wish to post three specific warnings.

- Teachers and users of mathematics must not assume that our "important ideas of mathematics" can be conveyed to students simply by turning to standard textbooks and looking up the topics we cite. In particular, in every case the topic listed must be seen by TAs as emerging from and contributing to the cycle of mathematical processes described above.

• Every topic must be approached through, and then studied more deeply with, problems. The challenge of problems that are seen by students as interesting, significant, or going beyond predetermined rules, is an essential pathway to the multifold mathematical experience that must underlie a proper grasp of what mathematics is and what it can do. In this report, we give just an occasional problem to illustrate what can be done to involve students in some particular part of one of the subject areas. However, to achieve a proper curriculum, it is essential to develop such problems for *every* topic mentioned.

• We have indicated our belief that TAs should know and understand the items that we list below under the various subject headings. However, neither knowledge nor understanding is an on/off state; rather, there are many gradations of each. Some facts must be recognized and responded to instantly, without reflection; others should be retrievable with a well-practiced algorithm; for still others, retrieval after some trial and error in setting up an algorithm will suffice; and finally, there are some facts for which it suffices that they should be recognized as having been seen, and that the TA knows how to start locating them by using reference books. As to understanding, most significant mathematical ideas require a gradual and long-term maturation process before they can be well understood. Concepts must be revisited at various grade levels and in various contexts before they can function properly in the total mathematical experience of a TA.

ARITHMETIC

TAs should be able to receive and originate communications involving various kinds of numbers: counting numbers and zero, decimal numbers, simple fractions, and the negatives of these. TAs should be able to locate roughly the place of such numbers on a "number line," and hence, they should be able to compare numbers to find the larger or smaller of a pair.

TAs should be able to use numbers to solve a large variety of problems arising in everyday experience. In particular, TAs need to be able to obtain sums, products, differences, and quotients in a variety of ways and with variable accuracy, and they should be able to decide which ways and degrees of accuracy are reasonable in various contexts.

The ways of obtaining sums and products of whole numbers should include use of a calculator, use of standard algorithms, mental arithmetic, and rough estimation. Memorization of sums and products of one-digit numbers should be sufficiently thorough to produce instantaneous answers. This capability, in fact, is the basis of all further forms of computation. However, the need for precise sums and products for multidigit numbers without the use of a calculator would be so rare that there would be no need for a great deal of practice of standard algorithms to achieve quick execution.

Indeed, because estimation is used frequently to monitor the use of calculators, algorithms that are *now* traditional may well be replaced by different algorithms that involve successive refinement of approximations, when computation by hand is required. For example, the procedure of "adding from the right" can be replaced by "repeated adding from the left," which would often be left uncompleted when full precision is not required. To illustrate: 327 + 485 would be first approximated as 300 + 400 = 700, then refined as 700 + (20 + 80) = 800, and finally made exact as 800 + (7 + 5) = 812.

Mental arithmetic involves the active use of "laws," or identities, that hold for the arithmetic operations (commutative, associative, and distributive). To illustrate: An easy way to compute 97 × 32 mentally is (100 − 3) × 32 = 3,200 − 96 = 3,104.

Estimation includes having some sense of the relation between two very large or very small numbers. For instance, if the national debt increases by about $300,000,000,000 this year, and the current population is about 220,000,000, about how much will each person's share of the debt increase? To handle such large numbers efficiently, TAs should be able to convert them into the form 3×10^{11} and 2.2×10^8 (and similarly to express very small numbers by using 10 with negative exponents), and to obtain products and quotients by using numbers expressed in this form.

As numbers are put to work in various problems, TAs should be able to shift scales by changing units. For example, how many meters per minute are traversed by a car going 100 kilometers per hour?

TAs should understand that simple graphs may be useful in making approximate calculations. If we have a large batch of numbers to be multiplied by 1.7, a straight line through the origin (of a coordinate system) and the point (10, 17) will enable us to estimate each of these products. Used in another way, the same line can be used to approximate the results of dividing many numbers given by 1.7.

Ratios and proportions are natural tools to use in a vast range of problems. Although numbers in decimal form are the most common way to express these, fractions with one-digit denominators are very natural tools for forming intuitive judgments and getting rough estimates. In order to compare, add, and multiply these, we have to be able to handle fractions with two-digit denominators. TAs should be able to express a pair of fractions with one-digit denominators as a pair of equivalent fractions having a common denominator.

We have been writing of ways to compute with numbers, but the big task for TAs is to know *when* to perform the various arithmetic operations. In traditional textbooks, this task is posed by various "word problems," many of which appear artificial and outside the normal range of interests of the students. By the year 2061, TAs should have come through many years of increasingly complex and diverse activities in which counting and measurements have naturally led to numbers that have had to be combined arithmetically to solve problems of interest. Among such activities will be shopping, travel, watching sports events, cooking, investing money, gambling, and playing games. To illustrate: A woman ran the first quarter-mile in 68 seconds, so what will her time be for the mile if she keeps up that pace?)

In the earliest years, these number-related activities will arise in play, but students' interests can easily be led to problems of an oconomic character involving production and exchange of goods that are valued by children. The beginnings of the acquisition of information and ideas that are the basis of the physical, biological, and behavioral sciences can be well started in the elementary school years, and these will provide a richer and more natural domain in which the exercise of arithmetic skills can be located. As students get an increasingly wide experience with real-world operations whose effects can be predicted by adding, multiplying, subtracting, or dividing numbers of various kinds, their understanding of the mathematical operations themselves will deepen, and their facility in employing arithmetic as an applied science will increase.

ALGEBRA

TAs should have a clear understanding of the concept of a solution of a system of equations: They should be able to check whether a proposed solution is in fact a solution. They should understand that a system may have no solution, exactly one solution, or many solutions: A set of requirements may be inconsistent, may define a unique course of action, or may allow several options.

TAs should understand arithmetic and geometric progressions in relation to simple and compound interest, and they should have seen formulas for the sum of arithmetic and geometric series.

Whenever possible, TAs should be familiar with pictorial representations of algebraic ideas. Here are some examples:

The following equation

$$(a + b)(c + d) = ac + ad + bc + bd$$

is represented by

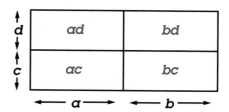

The successive numbers in an arithmetic progression are equidistant on the number line

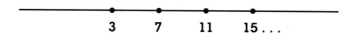

The statement

$$1 + 2 + \ldots + n = n(n + 1)/2$$

is represented by

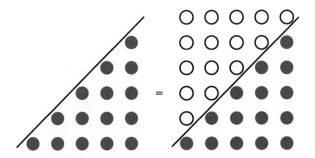

The solution of a linear equation
$$3x - 1 = 7 - x$$
is represented by the x coordinate of the intersection of two lines:

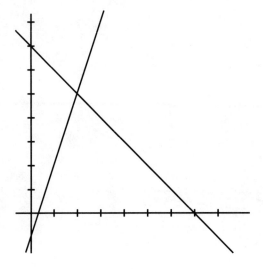

A quadratic equation may have two roots or one root or no roots:

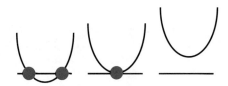

A product set is represented by a rectangular array: $\{A, B, C\} \times \{D, E\}$ is represented by

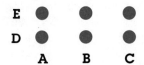

Product sets, or permutations, may be represented by paths or trees:

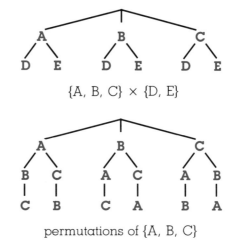

{A, B, C} × {D, E}

permutations of {A, B, C}

Addition of vectors is represented by the parallelogram law:

$$(a, b) + (c, d) = (a + c, b + d)$$

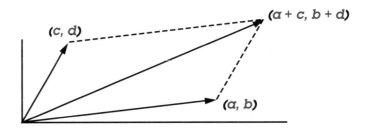

GEOMETRY

TAs should be familiar with basic geometric objects: point, line, angle, triangle, rectangle, square, polygon, circle, plane, cylinder, cone, sphere, and box. They should also be familiar with the relations *perpendicular* (two lines, line and plane, and two planes) and *parallel* (two lines, line and plane, and two planes).

It is important that TAs know something of the history of geometry, including that of the parallel postulate and the three classical construction problems: circle squaring, angle trisection, and cube duplication.

Put on an irregularly shaped lot with a tape measure, TAs should be able to make a scale drawing of the lot and then estimate the area and perimeter of the lot from the scale drawing. They should be able to calculate the surface area and volume of a box or cylinder, and should be able to estimate the surface area and volume of an irregularly shaped object, such as a house. They should have an intuitive understanding of the concept of similar shapes, and should also know that area varies as the square, and volume as the cube, of linear dimensions.

TAs should understand symmetry of an object as invariance under certain transformations, as in the following examples:

- The propeller is invariant under a rotation of 120 degrees:

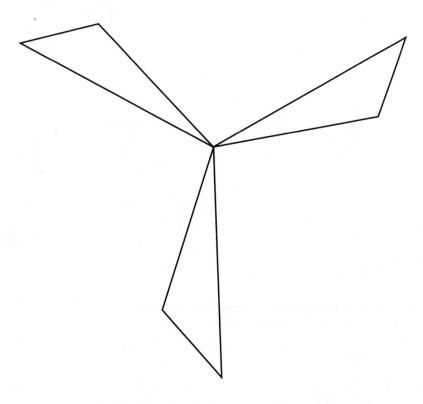

- The gingerbread man is invariant under reflection through its central axis; that is, it has bilateral symmetry:

- The infinite ladder is invariant under (a) a horizontal shift of one foot:

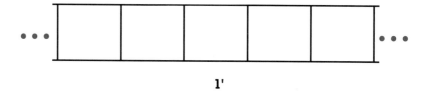

1'

The infinite ladder is also invariant under rotation through 180 degrees about the midpoint of any rung, reflection through any rung, reflection through any line parallel to a rung and halfway between two rungs, and reflection through the line that joins the midpoints of the rungs. (The last two conditions actually follow from the first three given above.)

TAs should know the Pythagorean theorem, and should be able to calculate the distance between two points in three-dimensional space from the coordinates of the points. It is probably not important for them to know many of the theorems of plane geometry, such as that the angle bisectors of a triangle are concurrent or that an angle inscribed in a circle is measured by half its arc. However, they should have followed a few proofs of such theorems, just to get the idea that proofs are instructive and not merely demonstrations of the obvious.

ANALYSIS

The description of the square root, addition, and y^x keys on a calculator as *function* keys will be familiar to TAs. They should understand the concept of function as a rule that associates each object in a given set with a corresponding object in another set. They should be familiar with ways of defining functions, including formulas, programs (as in computer programs), and algorithms. The concept of function is absolutely central in analysis, and permeates all mathematics.

TAs should be familiar with standard functions and notation. For real-valued functions of one variable, they should understand the concepts (if not the words) "monotone," "bounded," "continuous," and "nonnegative." They should also understand the concepts "maximization of functions" and "minimization of functions," and know how to distinguish clearly between the input value that maximizes the function and the output of the function for that input. In addition, it is important that TAs be able to interpret graphs of functions.

TAs should understand the derivative of a function—that is, the slope of its graph—as the rate of change of the output in relation to the input change. They need not know a formula for the rate of change of the function x^2, but should be able, from the graph of a function, to draw the approximate graph of the derivative of the function and the approximate graph, starting at a given point, of a function whose derivative is the given function. Thus, if this graph

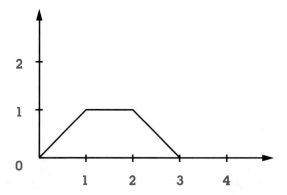

represents the position on the number line of a point as a function of time, TAs should be able to graph the velocity of the point as a function of time. And if the graph represents velocity as a function of time, TAs should be able to graph position, for a given starting point, as a function of time.

For real-valued functions of two variables, TAs should know that the graph is a surface. They should be able to interpret level curves, as seen in newspaper weather maps, or parallel sections (slices). They should understand that such curves and sections give a method for "visualizing" objects of higher dimension.

TAs should understand the meaning of a simple differential equation such as $y' = \lambda y$ (and should have encountered this equation in studying compound interest, population growth, and radioactive decay) or a simple system such as

$$dy/dt = y(x - 1)$$
$$dx/dt = x(2 - y)$$

(TAs may have encountered such systems in studying predator/prey relationships, and should be able to calculate approximate solutions, using difference equations.)

The differential equation $y'' = -y$ describes, approximately, the motion of the bob of a pendulum. TAs should be able to calculate—that is, program a computer to sketch—approximate solutions for different initial positions and velocities, and to verify (and check experimentally) that the period is approximately independent of initial conditions.

DISCRETE MATHEMATICS

You have 300 cards, each with a name on it, and must put the cards in alphabetical order. Method 1 first sorts the cards into, say, 10 piles—one containing names beginning with A or B, the second containing names beginning with C, D, or E, etc.—and then alphabetizes each pile. Method 2 divides the cards into, say, 10 roughly equal piles, alphabetizes each pile, and then merges the 10 alphabetized piles. Which method is faster? Is 10 about the right number? This problem, suitably formalized and made much more precise, is an example of a problem in discrete mathematics, an area that computers have made much more important. Some topics in discrete mathematics that TAs should have some familiarity with are algorithms (a precise scheme for alphabetizing a set of cards is an example of an algorithm), truth tables, elementary combinatorics, optimization, and tree diagrams. The following are three more examples.

Turing Machine

In the coming decades, TAs will be surrounded by computers. A Turing machine is a simple conceptual computer, and all real computers can probably be modeled by Turing machines. TAs should be familiar with a few Turing machines so as to have a general conceptual understanding of the capabilities and limitations of real computers.

A Turing machine can be thought of as a machine with a finite set of states, which can read 0's and 1's from a tape:

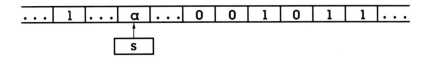

It has a finite list of instructions, each of the form "if you are in state s looking at symbol a, then do x and move to state t", where x is one of four operations: (0): put 0 for a, (1): put 1 for a, (L): look at the symbol to the left of a, and (R): look at the symbol to the right of a.

Here is an example—a machine with four states (0, 1, 2, 3) and five instructions:

s	a	x	t
0	1	0	1
1	0	R	2
2	1	0	3
3	0	R	0
2	0	1	1

TAs should be able to verify that this machine, started at the leftmost 1 of a tape that has n consecutive 1's and the rest 0's, will finally halt (reach a stage where it has no instructions—that is, in state 0 looking at a 0) with either 0 or 1 on the tape, according to whether n is even or odd: It computes the function $f(n) = n \mod 2$.

Games

TAs will probably be interested in games. Accordingly, they should understand the powerful concept of *strategy:* a function that assigns to each position in a game such as chess or backgammon a corresponding move, thus reducing the game to one in which each player has only one move (the choice of a strategy). They should understand the zero-sum concept, the use of mixed (randomized) strategies, and the notion of expected value, and they should be familiar with the ideas expressed in the von Neumann minimax theorem, which assigns a value to each zero-sum, two-person game.

TAs should recall having studied the game called Prisoner's Dilemma, a two-person game in which each player, by behaving so as to try to maximize his individual income, will get a certain amount—say, 1 unit—but in which, if each player behaves less rationally and more cooperatively, each will get a larger amount—say, 3 units. The arms race and many other real-world conflicts have aspects that resemble Prisoner's Dilemma.

TAs should be familiar with games that have both competitive and cooperative aspects, such as the three-person game in which, if A and B can agree on how to divide a dollar, C must give it to them. (OPEC operates a bit like this.)

Social Choice

As citizens who vote, TAs should be familiar with the concept of a *social choice function*, which acts on the individual preferences of the members of a group and produces as its value the group preferences. TAs should recall having studied the impossibility theorem proved by Kenneth Arrow—a theorem that asserts that certain plausibly desirable properties of social choice functions are mutually inconsistent.

LOGIC AND SET THEORY

TAs should certainly know how to reason—and should use that skill constantly. In its most basic form, reasoning involves using the words "not," "and," "or," "if . . . then" in a mathematically precise way. The mathematical meanings of these words are expressed by truth tables. TAs should understand how to use such tables to determine the truth value of a complex sentence that uses these connective words, on the basis of the truth values of its component sentences.

Without resorting to the use of tables, TAs should know how to simplify complex sentences by using various equivalencies of form. For example, not-(p and q) $\equiv$ (not-p or not-q), where we use $\equiv$ for equivalence; (if p then q) $\equiv$ (if not-q then not-p). However, TAs should also know that pairs of sentences that look similar may *not* be equivalent; for example, (if p then q) $\not\equiv$ (if q then p).

When we have equivalent sentences, we can replace either one with the other in our reasoning. However, the word "and" allows us to replace two sentences, such as *today is hot* and *tomorrow will be cold*, with an equivalent single sentence, such as *today is hot and tomorrow will be cold*.

On the other hand, the relation between a pair of sentences p, q and the compound sentence p or q is *not* one of equivalence. In the mathematical sense, we can infer the truth of p or q from knowing the truth of the single component sentence p; but the two sentences are not equivalent, because p or q may be true even though p is false. We express the relationship between the sentences by saying that p implies p or q, but not conversely.

TAs should recognize various simple patterns for establishing implications. For example, [p or q, not-p] implies q.

Whenever one sentence, r, implies another one, s, then the sentence *if r then s* is logically true. Conversely, the pair of sentences [*if r then s*, r] implies the sentence s. However, TAs should know instantaneously that the pair [*if r then s*, s] does *not* imply r.

Two forms of reasoning are so powerful that TAs should be using them *often*. One is called proof by cases. In this form of reasoning, if we know the truth of p or q, then the pair [*if p then r*, *if q then r*] implies r. The other is called proof by contradiction. In this form of reasoning, the pair [*if p then q*, *if p then not-q*] implies *not-p*.

We stress that it is not enough for TAs to be able to recognize the correctness of such reasoning when others use it. They must be able to *use* such reasoning to solve problems in all domains of experience.

The equality sign, =, which students encounter in algebra, is actually as basic a logical tool as the connectives *not*, *and*, and *or* and can be useful in all realms of experience. The sentence *Mary = John's teacher* logically implies (the daughter of Mary) = (the daughter of John's teacher). Hence, the pair of sentences [*Mary = John's teacher*, *Mary's daughter is ill*] implies *the daughter of John's teacher is ill*.

TAs must also be able to deal with elements of the part of logic dealing with the words "every" and "some". They should be able to follow and produce reasoning illustrated in the following two examples.

- For every counting number, there is some counting number larger than it. Hence, there is no largest counting number: So it's *not* the case that there is some counting number that is larger than every other counting number. (On the other hand, there is a counting number that is less than every other counting number; so it's *not* the case that for every counting number there is a counting number less than it.)

- If 4 is a factor of some counting number, then 2 must also be a factor of it. That is, for *every* counting number, if 4 is a factor of it, then 2 is a factor of it. In particular, since 1 and 2 are counting numbers, we conclude that if 4 is a factor of 1, then 2 is a factor of 1; and if 4 is a factor of 2, then 2 is a factor of 2.

TAs should be used to considering all sorts of sets of things that come to their attention. Very few such sets have fixed names in our language, and TAs should instinctively use letters or other symbols to serve as temporary names for the sets of interest in connection with some aim or problem. They should come naturally to consider intersections, unions, and differences of such sets, and should be familiar with a standard notation for combining the names of the original sets to discuss the sets resulting from such operations—for example, $A \cap B$ for the intersection of the sets A and B.

TAs should also consider all sorts of (binary) relations that connect pairs of things. For instance, arithmetical relations such as < or =; human relations such as *father of*, or *neighbor of*; and geographical relations such as *east of* or *higher than*. TAs should realize that intersections and unions of relations lead to new relations, just as they do for sets; for example, (father of) ∪ (mother of) = (parent of). But there are new operations for relations that have no counterparts for general sets, such as *converse*; for example, converse (parent of) = (child of).

PROBABILITY AND STATISTICS

Uncertainty is everywhere—uncertainty about tomorrow's weather, next Monday's closing Dow-Jones average, the success of the next space shuttle, the winner of next year's Super Bowl. Uncertainty is modified by information. So TAs should know something about probability, which measures uncertainty, and about statistics, which treats both the collecting and the summarizing of information and the use of information to modify uncertainty. Here are some specific competencies TAs should have:

Descriptive Statistics

TAs should understand the concepts known as mean, percentile, standard deviation, and distribution of a variable, so that they can estimate them for variables with which they are familiar, such as the height of adult males, the weight of dogs, or the price of houses. They should be able to interpret the histogram of a variable (as seen in newspapers or on a computer screen), and estimate from it the mean, the median and other percentiles, and the standard deviation of the variable. They should be able to look at a scatter diagram for two variables and estimate the means and standard deviations of the two variables and their correlation. They should be able to distinguish clearly between correlation and causality.

Probability

TAs should understand that all probabilities are conditional, and that they change with information. (Question: A box contains two black balls and three white ones. The balls will be drawn one at a time at random, without replacement. What is the chance that the second ball drawn will be black? TAs should not respond, "It depends.") TAs should know Bayes' formula, which shows how to use evidence to modify uncertainty: posterior odds = prior odds × odds from evidence.

Sampling

TAs should understand random numbers, and what it means to draw a random sample from a population. Given a distribution, TAs should be able to draw a random sample from the distribution (easy when computers are used) and study the results. For instance, TAs should be able to take the distribution of prizes in the California lottery and, by repeated sampling, estimate the distribution of winnings over the next five years with a given strategy. Lacking enough information to specify the distribution, TAs should recognize this (and perhaps seek the information). They should be able to answer the question, "Suppose in a random sample of 1,000 California voters, 700 favor candidate A. Is this overwhelming evidence that a majority of all California voters favor candidate A? (Correct answer: "Yes.") Perhaps even more important, TAs should understand the importance of the word "random" in the above statement; they should also know that people who phone in answer to a newspaper request are not a random sample.

A SAMPLE PROBLEM

Mathematical problems do not come labeled "arithmetic," "algebra," "geometry," etc., and often require the use of ideas from several branches of mathematics. The following is an example of such a problem—and TAs should have solved similar problems.

The Problem

You have a supply of wheels, flats, rods, and seats that you can use to make scooters and tricycles.

Each scooter will need 2 wheels, 1 flat, 2 rods, and 0 seats.

Each tricycle will need 3 wheels, 1 flat, 4 rods, and 1 seat.

You have 30 wheels, 12 flats, 40 rods, and 8 seats.

(A) Can you make 8 scooters and 4 tricycles?

(B) Can you make 4 scooters and 8 tricycles?

(C) Mark the set of all points (x, y) in the plane with $x \geq 0$, $y \geq 0$ for which you can make x scooters and y tricycles.

(D) If you can sell each scooter for $100 and each tricycle for $200, how many of each should you make to maximize your total income?

Outline of Solution

To make x scooters and y tricycles:

You will need $2x + 3y$ wheels, so $2x + 3y \leq 30$

You will need $x + y$ flats, so $x + y \leq 12$

You will need $2x + 4y$ rods, so $2x + 4y \leq 40$

You will need y seats, so $y \geq 8$

The answers:

(A) Yes; all inequalities are satisfied.

(B) No; $2 \cdot 4 + 3 \cdot 8 = 32 > 30$: not enough wheels.

(C) The possible points are marked with a dot (●) in the following diagram.

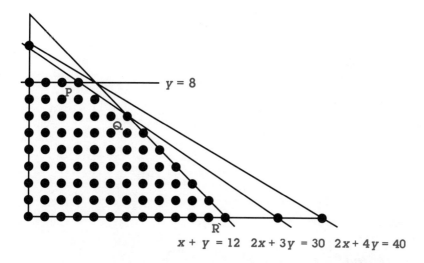

$x + y = 12 \quad 2x + 3y = 30 \quad 2x + 4y = 40$

(D) One of the three points $P = (3, 8)$, $Q = (6, 6)$, $R = (12, 0)$ is optimal. Their yields are $300 + 1{,}600$, $600 + 1{,}200$, $1{,}200 + 0$, so P is best. You should make 3 scooters and 8 tricycles worth a total of $1,900.

Comments

To solve this problem, TAs must translate simple ideas into mathematical language. They must also plot points, recognize the equation of a line and graph the line from its equation, know that a linear inequality represents a half-space, translate simultaneous conditions into intersections of sets, solve a system of two linear equations in two unknowns, and understand the concept of maximizing a function over a set. They will not necessarily see that the maximum must occur at P, Q, or R, but they should see that it must occur at one of the 10 dots that have no dots to the north or the east, since moving north or east increases one's income. The actual maximization process, which is extremely important in the many large-scale real-life applications of linear programming, is less important here than the clear mathematical formulation of the problem.

TAs should also be able to formulate the extension of the problem in which unicycles and wagons can also be made. It becomes a problem of maximizing a linear function of four variables over a region of four-dimensional space determined by linear constraints. TAs should understand that four-dimensional space, although perhaps not directly geometrically visualizable, is not mysterious, and is a practically useful concept—here the idealization of the possible production vectors for unicycles, scooters, tricycles, and wagons.

SECTION 4

MATHEMATICS, SCIENCE, AND TECHNOLOGY

Not only particular mathematical problems but also whole areas of mathematics have developed historically from one or another of the empirical sciences. For example, the subject of differential equations clearly grew out of mathematical efforts to solve problems of physics and astronomy. In the other direction, new methods of dealing with areas of an empirical science sometimes originate in mathematical ideas that were developed earlier for quite different purposes. For example, the direction of important kinds of research in the field of economics was radically changed by the impact of game theory and the related method of linear programming.

In school curricula, the presentation of material on science and technology has been largely descriptive, with very little mathematical formulation or analysis. This situation diminishes the opportunity for students to appreciate mathematics as well as science and technology. Such opportunities must be developed and strengthened by future mathematics and science curriculum specialists working together. Ideally, the interaction of mathematics, science, and technology will emerge in the future curriculum from hands-on experience generated by students' interests under the guidance of expert teachers. In this report, we do not attempt to provide examples, for our knowledge of requisite material is very limited and any choice of a small number of examples would necessarily be unrepresentative. That would leave readers with a distorted idea of the kinds of curricular innovations that we would like to see.

The above discussion focuses on the interaction of mathematics with traditional areas of science and technology, but the relation between mathematics and the use of computers deserves special consideration in the design of future school curricula. The penetration of computers into businesses, schools, and homes across the nation resulted from discoveries in mathematical logic in the mid-1930s, application of those discoveries to problems of physics in the military effort of World War II, and subsequent investment in technological development to make the application of those discoveries economically feasible in a broad range of business and governmental activities.

There is no doubt that the spread of computers in our national life will require various forms of computer-related education in the schools, but not all of this will be related to mathematics instruction. Still, the interaction of computer use and mathematics learning is so great that there will be a good many changes in the school mathematics curriculum as a result. The following is a survey of some of the forms of change that we can expect.

- Computers will increasingly relieve teachers of aspects of pedagogy that can be routinized. For example, computers will be able to feed practice problems to students at their individual levels, diagnose error patterns, and provide students with ego-free opportunities to experiment. Thus freed, teachers will be able to devote more time to leading students into new areas of study.

• By relieving students of heavy burdens of computation, computers will make it possible for them to take up new areas of mathematics at earlier grade levels. For example, the analysis of data and the drawing of probable inferences from them involve mathematical ideas that could and should be taught as early as junior high school years. That is currently avoided because the large amounts of arithmetic met in examples use up too much of the students' time and attention.

• As indicated earlier, presentation in school of material from the physical, biological, and social sciences tends to be descriptive rather than mathematical. Part of the reason is that to use mathematical methods properly would require lots of applications that generate heavy loads of arithmetic. When those loads can be shifted to the computer, students will be able to dig into the relation of theory to empirical phenomena.

• Mathematical subjects heretofore omitted from school curricula because they were not needed for later work in mathematics or other studies will become necessary to enable students to understand how they will be called on to use computers. Combinatorics, the theory of relational graphs, and parts of logic are some of the areas given a new importance by computers.

• Other mathematical areas already treated in the school curriculum will require a much heavier emphasis because of their extensive use of computers, whereas still other areas of traditional emphasis will be treated lightly because of the extent to which computers will take over their function. For example, estimation will become a much higher priority and algebraic simplification will require much less drill.

• In the study of algorithms, there will be a great shift away from memorizing "the standard way" toward making various kinds of calculations, and toward concern with the relative speed and efficiency of various computational methods for each kind of calculation.

• The large amount of guessing and checking that precedes the formulation of a proposition for which a proof will be sought can be shortened with the help of computers. This change will help bring the exercise of proving into new parts of the school mathematics curriculum.

In this report, we are not providing detailed examples of material that will ebb from and flow into the mainstream mathematics curriculum to illustrate the above types of change, because interpenetration of computers and school mathematics is still at an early stage and specific examples might prove more confining than enlightening. There is no doubt, however, that in the later phases of Project 2061, it will be necessary to provide substantial indications of how the mathematics curriculum should respond to the changed environment that computers are creating.

SECTION 5

MATHEMATICS AND LANGUAGE

Much of the difficulty that students now experience in seeking to learn mathematics is due to their awkwardness in handling the language used in formulating the subject, rather than to any intrinsic complexity of the concepts involved. Looking ahead to the time when TAs—typical adults without mathematical study beyond the age of 18—will want to express, as well as understand, mathematical ideas used in the context of many domains of experience, it is essential that they learn to use the language of mathematics at an early age and retain control of its use throughout their lives.

When we speak of mathematical language here, we do not have in mind an elaborate symbolism replete with formulas full of esoteric characters. Rather, we mean the careful use of natural language, clarified by certain conventions that eliminate ambiguity, and supplemented by the use of variables and carefully defined terms. These features of mathematical language enable mathematicians to formulate their concepts with utmost precision and to communicate propositions and their proofs in a mode that carries complete conviction.

The conventions that eliminate ambiguity are principally those that confine the disjunctive connective "or" to its inclusive sense, and that fix the meaning of the conditional connective "if . . . then" in the manner of logical truth-tables (so that a sentence of form "if P then Q" is considered true in every case except where P is true and Q is false).

Mathematicians are also completely careful about the order in which negation and generality are employed in the same context. To deny with mathematical precision that all men will wear ties, one must say "not all men will wear ties." The sentence "all men will not wear ties," although occasionally used with intent to express the same sense, means for the mathematician that not a single tie will be worn by any man. (The sentence "not all men will wear ties" is equivalent to "some men will not wear ties.")

It is the free use of variables and special definitions to supplement natural language that gives mathematical language its creative power and enables it to be used for the most subtle analysis of ideas. Although the variables that appear in high school algebra are sometimes treated by students as meaningless marks to be moved about by rules that seem to them to be arbitrary, mathematicians give meaning to a variable each time it is employed. It is true that the same symbol—usually a letter of the alphabet—changes its meaning from one context to another, but this is consistent with its designation as a *variable*. Although the meaning of "b" or "x" changes, the changes are completely controlled by the mathematicians who are using them.

Variables are by no means confined to numerical discourse; they can refer to objects of any kind whatever. In their simplest use, they serve temporarily as names for objects that have no proper name or simple descriptive identification in our unaugmented natural language. For instance, if we are looking at a pile of buttons, we might point to one of them and say, "Let's call this one b. How much do you think b weighs?"

More characteristically, a variable will be used as the name for one of a bunch of things without specifying *which* one; typically, the word "any" is used in such a case. For instance, we might say, as we look at the pile of buttons imagined above, "Let c be any button in this pile." If we see that no two buttons of the pile have the same size or color and that the smallest button is black, we could properly introduce a second variable by specifying "If c is not black, let d be the largest button smaller than c; otherwise let d be the black button." Having agreed on the meaning of the letters "c" and "d" in this context, we could start drawing mathematical conclusions. For example, we can conclude: If c @ d and d is black, then the only buttons in the pile that are not larger than c are d and c itself.

Even if a particular object has a proper name or can be specified by a precise descriptive phrase, the mathematician may introduce a variable as a temporary name for it simply as a means of economizing. For instance, if we have quite a few things to communicate about, *the longest river in North America*, we may prefer not to repeat that six-word description of the river for every reference; even its proper name, "Mississippi River," might seem too long. So we could begin by saying, "Let r be the Mississippi River"—and we could then give details: r originates in Minnesota; the Missouri River flows into r; Mark Twain wrote about r; etc.

The impulse to economize by abbreviated notation is a powerful one in mathematics. It often leads to an extended series of definitions, in each of which a single symbol (or short term) is introduced as equivalent to a descriptive phrase that contains several component symbols introduced by earlier definitions. Such linguistic patterns enable the mathematician to chew intellectually on ideas that would remain indigestible by the intellect if expressed at length in the original terminology of unadorned natural language.

We have indicated that variables may stand for buttons or rivers, as well as for numbers, but it is important to mention explicitly that they can also be used to refer to *relations* that connect different kinds of things. For instance, we may consider the relation "M" that holds between a person, "p," and a river, "R," just in case "p" is the mayor of a town situated on the river "R." The relation "M" might be of interest to a business that manufactures certain kinds of waste-disposal equipment, and its engineers might find it useful to use the letter "M" with this meaning in their analysis.

We have given examples in which letters are the symbols used as variables, but symbols normally considered constants, such as the familiar plus sign of arithmetic, may be used as variables in some parts of mathematics. This happens because mathematicians encounter and construct systems that have an operation that *resembles* the addition of numbers; for example, the new operation may act on pairs of vectors instead of on pairs of numbers. It turns out that the vector operation obeys principles familiar to us from our study of the addition of numbers, such as the associative law. To make use of the skills we've developed in handling the plus sign, which are based on such laws, we can use the same symbol for the operation on vectors, and call it vector-addition. We may even use the plus sign several times in the same sentence, sometimes referring to numerical addition and other times referring to vector addition; we rely on the symbolic context of each use to inform the reader as to which meaning is intended. The beginning student, not properly forewarned, may find this utterly confusing. Nevertheless, the linguistic skill required to master such use is far less than children exhibit when they learn how to conjugate irregular verbs at ages 4 and 5. Once made a part of an individual's naturally used language, these aspects of mathematical language enormously enhance the individual's capacity to carry out mathematically precise discourse and analysis in a broad range of experience.

In addition to variables, which may serve as temporary names for mathematical objects, and special symbols such as +, which serves as a permanent name for the mathematical operation of addition, mathematicians use ordinary words as names for the objects of their discourse. Sometimes, the mathematical meaning of such a word is very different from its normal meaning; for instance, the mathematical term "ring" refers to a kind of number system (of which the system of integers is one example). Such usage may at first appear disconcerting, but once learned it creates no difficulty or confusion.

At other times, however, the mathematician's use of a word may closely resemble ordinary use, yet it may differ slightly; unless all the users are sensitive to such differences, misunderstandings may occur. For example, in ordinary discourse, the word "circle" sometimes refers to a one-dimensional closed curve of the kind drawn by a compass; at other times, it refers to the two-dimensional region lying within such a curve. To distinguish these geometric objects, the mathematician confines the use of the word "circle" to the curve and uses the word "disk" to refer to the region. It may be that a failure to distinguish sharply between a two-dimensional region and its one-dimensional boundary is at the root of the confusion some people seem to have about the difference between area and perimeter. (This was a distinction that the new math made great—but unsuccessful—efforts to establish.)

Perhaps the most subtle alterations of language by mathematicians occur when a natural word is specialized *without* giving notice by means of a formal definition. In the theory of games, for example, it has become the practice of mathematicians to say that a player is pursuing a "rational" strategy if he or she has selected a strategy designed to maximize the payoff. However, as the Prisoner's Dilemma indicates (see Section 3 above), the normal use of "rational"—which is, of course, ambiguous—may well be in conflict with this particular usage.

When many specialized symbols have been brought into mathematical language by a long chain of definitions, whole sentences are often composed entirely of such symbols; equations between formulas are typical examples. Someone who has not worked through the definitions and practiced using sentences with such a heavy density of specialized sentences finds this part of mathematical language particularly mysterious—and may think that the task of understanding it appears hopeless. But really, it is much easier to bring oneself to the point of understanding such equations than it is to come to understand fairly simple sentences in an esoteric language, such as those used by some of the ancient Egyptians—even though children at one time could deal easily with such sentences. The essential element of linguistic mastery is the opportunity to use the language regularly, at a young age, in an environment in which expert users are available to be imitated and to give corrections. In such an environment the specialized terms, symbols, and rules of sentence formation become associated subconsciously with intuitive images and ideas that are related to nonlinguistic experience. Then, the real work of learning is no longer focused on symbolic manipulation but on developing intuitions associated with real-world processes.

In earlier centuries, and indeed well into the twentieth century, the inclusion of mathematics as a principal component of a liberal education was justified on the grounds that its study "taught one how to think." It was hoped that the study of mathematics would help people to analyze a broad range of life's problems with effective tools, even if mathematical formulas and theorems were not employed. Nowadays, that thesis has fallen into disuse.

It is quite possible, however, that the disrepute into which the old thesis—that mathematics teaches us how to think analytically—has fallen is due to the degeneration of mathematics instruction into drill on the memorization of standard computational algorithms. It may well be that the true potential for mathematics to strengthen general problem-solving abilities lies in the nature of mathematical language. In its precision, flexibility, and ease of use for representing new abstractions, and in lending itself to both deductive and computational invention, may lie the power that was classically thought to inhere in mathematical studies.

SECTION 6

EMOTIONS AND MATHEMATICS

What are the important ideas of mathematics that people should know and understand by the age of 18?" Our search for these ideas started with a reference to the fact that mathematics grows out of "ordinary human experience." So far, our analysis has led us to focus on mathematical processes, subject areas, relations with science and technology, and language. Now it is time to look at mathematics in relation to another facet of ordinary human experience: feelings. There is no way to come by ideas without experiencing feelings.

As in other fields, the concept of mathematical knowledge can be abstracted from the conditions under which it is acquired, and there is much of value to be gained by studying this abstraction. But as soon as we ask, "How do TAs—typical adults—learn what they know?"—and even more if we ask, "How come TAs don't know what they should?"—we must face the question of feelings. What people feel like doing has an enormous influence on what they do, on how they do it, and on why they don't do other things.

It is evident that for many people nowadays, the feelings engendered by their mathematical experience include fear or anxiety, or both. Across the country, in colleges and adult schools, courses such as Math Anxiety or Math Without Fear have appeared in recent years. Have you ever seen courses with similar names in which "Math" is replaced by one of the other commonly studied subjects—say, as in Music Anxiety or History Without Fear?

Why are many people anxious or fearful about mathematics? It is natural to seek an answer in what distinguishes the subject matter of mathematics from other subjects. However, if we do so too quickly, we may overlook an important source of human emotions: *other* people's emotions. All people, and especially young ones, have a sense of what others around them are feeling, and that tends to be reflected in their own feelings. People want to "belong," for that need is the foundation of any culture. If one grows up hearing parents and peers saying that "math is tough," that only unusual people like it, and that they themselves avoid it, one's feelings can be affected in a way that indelibly colors one's own contact with mathematics.

Incredibly, among the people who "infect" many children with math anxiety are some classroom teachers. Because elementary teachers must teach all subjects and quite a few have had math anxiety themselves, students may receive a sense of unease when it is time to change from another subject to math—even before the lesson begins. And at the secondary level too, because some teachers have been switched from their original specialty to teach mathematics (a subject that they may have been used to avoiding), their students may pick up negative feelings.

It is curious and important to notice that there is one part of mathematics—involving a very important and complex skill—

about which no one expresses fear or anxiety: counting. Because the language of counting is learned at the same time and in the same way as other basic parts of natural language, because counting emerges in the context of other aspects of experience that are important and interesting to children, and because counting enters into the social activities of children, a favorable emotional context generally surrounds the learning experience.

We remarked at the beginning of this report that the classroom experience of mathematics is now focused on memorization and practice in applying prepackaged algorithms, to the virtual exclusion of active participation in the processes of abstraction/representation, deduction, and application/comparison—and even the part of computation involved in constructing algorithms. The natural experiential context of mathematical activity being thus ruptured, a positive emotional environment for mathematics learning comes to depend on the enthusiasm of an exceptional teacher, parent, or friend, or indeed on a special talent that is excited by even a limited contact with mathematics. How widespread would the enjoyment of and success with mathematics be if a full range of mathematical processes were developed through interaction with other parts of children's activities? At present, the answer must be left as a matter of conjecture and, we hope, experiment.

Let us look more closely at the affective context in which the learning of mathematics takes place in most of our present-day classrooms. A module of experience generally begins either with the teacher introducing, explaining, and showing some topic or technique, or else with the teacher asking the children to work on some kind of problem based on material already presented. Why should students pay attention to the presentation? Why should they work at the problem? In a very large proportion of cases, the honest answer has to do with the fact that the teacher has been placed in a position of authority and can bring various forms of pressure to bear on students who do not work out the answer the teacher wants. By contrast, when students independently formulate a mathematical problem arising out of some project or activity in which they are engaged through self-motivated interest, their emotions of curiosity and eagerness combine naturally to move them into mathematical activity.

The emotional context is of great importance not only in *engaging* the wholehearted attention of students with particular mathematical problems but also in guiding them toward solutions. When seeking to meet a teacher's demand for quick production of memorized material, the student may experience a feeling of helplessness and frustration if the desired response does not come to mind upon demand; there's nothing for the student to do about it except look it up again. But if the student is seeking to find an answer to a self-set problem, there is no expectation that some particular prepackaged technique must be instantaneously brought forth. On the contrary, trial and error, following hunches, and experimenting—all the methods familiar to mathematicians and scientists—fall naturally into place. And when a successful answer is found to a self-set problem, the student is rewarded by a feeling of elation and the excitement of discovery in place of the rewards that too often are all that one sees in some classrooms—relief from the teacher's pressure or a grade of A.

We have been writing about emotions connected with the serious pursuit of mathematics, but there is a long tradition of undertaking mathematical activity as a form of relaxation. This tradition ranges from individuals working on puzzles and problems that may require protracted effort—thereby relieving the mind of preoccupation with some tension-producing problem—to forms of group play that may involve cooperative or competitive games. When such games are undertaken in the classroom, the positive emotions that are characteristic of children's play may lead to mathematical learning that remains alive, as compared with learning that is forced on students by a teacher's pressure.

To conclude our account of the emotional context of doing and learning mathematics, we wish to stress the social nature of those activities. The stereotype of the mathematician as a strange individual who works alone is belied by the increasingly frequent collaborations in mathematical writings, as well as by the common professional interactions of mathematicians with scientists and engineers. In the most successful classrooms, students talking among themselves about how to deal with mathematical problems provides a common setting for learning. When such conversations take place *outside* the classroom, as part of a larger shared experience, the pleasure of interpersonal participation lends a warmth to mathematical activity that transcends individual satisfactions.

And yet, the individual maintains an identity that passes through social interactions and continues. Thus, the emotional differences among individuals are a persistent factor that a teacher must not overlook in promoting learning—in mathematics as in other fields of study. The wise teacher will seek to reach each pupil in a special way, building on the interests and feelings of the individual.

SECTION 7

CONCLUDING REMARKS

We have now set forth the main parts of our answer to the question of what important ideas of mathematics everyone should know and understand, and it remains only to say something about how the parts are supposed to fit together and what is to be done with the result.

- Our overarching theme is that mathematics is a part of human experience. It emerges from everyday experience and can be reflected back on that experience.

- Mathematical activity consists of an interactive cycling of basic processes (Section 2) that lead from nonmathematical domains such as those of science and technology (Section 4) into the several subject areas of mathematics (Section 3), from one of those areas to others, and from one part of a subject area to other parts. Students should begin by getting practice in using the processes to increase their understanding of the external subject matter that is receiving mathematical treatment. However, in the end, they should come to understand the processes themselves, and their interaction, as fundamental components of mathematics.

- Adaptations of natural language are needed to communicate successfully about mathematical ideas. Facility in the use of such "mathematical language" should be developed in parallel with the earliest mathematical experience (similar to the development of other parts of natural language), and then should be strengthened by continuing use at all grade levels.

- Mathematical learning should be integrated with play. Many mathematical ideas should emerge from a variety of constructions and other projects having physical, chemical, and biological elements, as well as from games possessing economic and strategic characteristics.

- From the different kinds of activities mentioned in the previous paragraph, children should be helped to develop intuitive ideas about "how things work" in various realms of experience. They should learn how to "translate" their intuitions into hypotheses about mathematical models of the real-world phenomena, and they should get used to adjusting intuitions and models to fit with experience.

- Generation of the cycle of processes occurs when problems are formulated—problems that can arise outside or inside of mathematics. Currently, there is a strong effort by mathematics educators to develop challenging problems for classroom use, and to get teachers to use them. But in fact, the formulation of problems is an important part of mathematics in which students *themselves* ought to gain experience. Ideally, students at all levels should be helped to develop mathematical concepts and problems on their own, in the context of hands-on projects that become increasingly sophisticated at higher grade levels.

- The mathematical ideas set forth in this report are to be transformed into curricular ideas by a suitably constituted panel as part of Phase 2 of Project 2061. Among the tasks this panel must face is the determination of the different degrees and levels of knowledge and understanding (see the third warning—about knowledge and understanding not being on/off states—presented in Section 3) that it is appropriate to seek for the various mathematical ideas that we have mentioned in Sections 3 and 4. Ultimately, if our ideas are to have impact on mathematics learning, they must be implemented by teachers or other educational specialists. What kinds of skills and knowledge will they require, in addition to those presently required of teachers? It is clear to us that from the earliest grade levels on up, it will be essential not only that teachers know about developmental psychology and learning theories, but that they have a deep knowledge of mathematics in its interaction with real-world phenomena as well as of the way the fundamental mathematical processes function in many mathematical subject areas. Teachers will need to be sensitive to the emotional and linguistic dimensions of mathematics learning and will need to have the competence to help their students make progress in those aspects of mathematical experience, as well as in the intuitive and cognitive aspects. The development of a profession of such teachers presents a major challenge to our society, but it is one that must be met if we are really to provide citizens with the mathematical knowledge and understanding that they will need in the coming century.

POSTSCRIPT

Individual portions of our report elicited support or criticism in a varying pattern from the readers of the successive drafts. But the report as a whole generated one question that was clearly the most widespread reaction: How can we possibly expect substantially all 18-year-olds to learn and understand so much mathematics of the kind described in the report in the coming century when we see so much failure today at what seem to be much simpler parts of mathematics? Many students now elect to discontinue mathematics very early in their high school years. In many cases, this is the result of failure in remedial courses that seek only to strengthen the students' skills in handling elementary school topics.

Unexpectedly, the motion picture industry has recently presented evidence that the potential for learning mathematics may be far greater than is indicated by expectations based on past performance. In the spring of 1988, the film *Stand and Deliver* was released nationally. It tells the true story of a teacher at Garfield High School, located in the barrios of Los Angeles. This teacher, Jaime Escalante, proposed in 1979 to work with a class of ninth-grade students taking General Mathematics and to get them ready to pass the advanced-placement calculus course in their senior year. No student at Garfield had ever taken an advanced-placement exam. The school, with a predominantly Hispanic population, was set in an economically depressed area; serious problems of discipline and behavior were rampant in the school; and families wanted students to contribute to family income rather than spend time with their books.

Escalante's proposal was opposed by his colleagues and challenged by the students themselves. When the school principal gave Escalante permission to try, the chair of the mathematics department resigned, on the grounds that the students would surely fail and their self-confidence would be irreparably undermined. Undeterred, Escalante started to work with the students. Using unusual teaching methods, he transformed the class into a determined unit with the common goal of learning calculus.

In the spring of 1982, 18 students at Garfield took the advanced-placement examination. Subsequently, all received notice from the Educational Testing Service (ETS), which prepares and scores all such exams, that they had passed. The class and the whole school and community that had watched their tremendous effort were jubilant. Then, a few weeks later, ETS sent a notice to the principal saying that it had reason to believe there had been cheating and that the passing grades would not be certified to colleges.

There was consternation in the Garfield community. ETS sent a team of investigators to talk to the students, to find out what methods of cheating had been used. When Escalante asked for the evidence that there had been cheating, he was told that no other school had had *all* of its students who took the exam achieve a passing grade. Furthermore, since this was the first time any Garfield High students had taken the exam, it was simply unbelievable that all could have passed without cheating. When its investigators failed to obtain any admissions of guilt from the Garfield students, ETS proposed that the students retake the exam (another version of it).

In the end, a dozen of the students agreed to take the new exam. Within a few days, a special team from ETS headquarters appeared at Garfield High and personally administered the exam. All of the students passed. Only a few got the minimum passing grade of 3; there were quite a few with the maximum grade of 5; most got 4.

A scene of the principal receiving the scores of the students ends the motion picture story. But before the film credits are shown on the screen, a few statistics about Garfield High are reported. In every year after 1982, a larger number of Garfield students passed the advanced-placement calculus exam, growing from 18 in 1982 to a total of 87 in 1987. In 1987, Garfield High produced more students passing this exam than all but three other public high schools in the United States.

This story is not fiction, it is not speculation; it is history. The 1982 decision of ETS not to certify the original 18 passing scores was widely reported in the press at the time. The sequel shows definitively that it is possible, by suitable changes in the learning environment, to improve the mathematics learning of a school population far beyond the general expectations.

When we contemplate the growing stream of Garfield students who have made the transition from a terminal remedial math course in ninth grade to a college-level calculus course in twelfth grade, we are compelled to treat seriously the question of whether essentially *all* students can far transcend their present levels of achievement.

What elements in our panel report bear on this question?

The section on emotions and mathematics is central. It is clear from the story of Escalante's achievement at Garfield that his main accomplishment with the initial group of 18 students was to create a communal will to work toward the goal of passing the calculus exam and to instill a belief that they could succeed. As we have said, how people *feel* about their experience with mathematics is the key to what they can accomplish with it. If children see that they can use it to enhance their other realms of experience, if they see that they can succeed in doing what people around them are interested in doing, and if they can feel the pleasure of interacting with their peers in group projects, their attitudes toward mathematics will be totally different from those that now prevail in the schools.

Our report describes mathematical ideas that everyone ought to know by the age of 18. How would such ideas contribute to generating the feelings just described, which motivate students to master much more mathematics than the current norms?

In discussing the *processes* of mathematics, we have emphasized the idea that mathematics arises from and feeds back into a wide range of normal experience—including kinds of experience, such as play, that are known to be very attractive to children. We must recognize, however, that we have channeled a vast number of students into mathematics avoidance. We have done this by emphasizing the rote acquisition of standard algorithms used in arithmetical computation and by withholding from children the experience of formulating mathematical problems related to their own interests, shaping abstractions, making and checking conjectures, moving among ideas by deduction, and trying their own applications of mathematics and learning to evaluate them.

Now for a crucial question. Let us suppose that by teachers' turning the study of mathematics to the subject of processes relating mathematics to other realms of experience, students become much more interested in their classwork than they are at present. Let us also suppose that this interest leads them to generate the kind of group dynamics and personal sense of accomplishment that led the Garfield High students to their success. The question: Is there some sort of special ability that will be needed to achieve understanding of the kind called for in our report, or is there some tool that we can give to all students that will be adequate for the task?

It is our contention that there is a specialized pattern of using and supplementing natural language that is essential to the accurate expression and exchange of mathematical ideas. Most children never have a chance to hear, imitate, and finally use this language during the great early years of language acquisition. Certainly, youngsters who learn a second language by speaking it in a family or community context achieve far more than those who study the second language only in a classroom situation where communication is in the primary language. Is it unreasonable, then, to expect that there might be a *great* increase in mathematics learning if we could somehow provide the needed linguistic environment at the right time?

We have been arguing, above, that the kinds of mathematics that the panel report says everyone should know by the age of 18 may in fact be within reach of everyone. The argument begins by considering the students at Garfield High, described in the film *Stand and Deliver*, who achieved far more than people thought they could, in view of their past accomplishments. Referring to parts of our report relating to the emotional context of mathematics, and to the relation of mathematics to the other realms of experience in which children have a primary interest, we have suggested that these factors could bring about a vast improvement in motivation for *all* students, perhaps even greater than was achieved at Garfield High. Finally, by emphasizing the close relation of mathematical language to the content of mathematics, we have suggested that a serious cultivation of mathematico-linguistic tools for children could provide the requisite tools for children to acquire the same degree of mastery of mathematics as of their native language.

However, if high mathematics potential is indeed ubiquitous and almost everyone is now underachieving, who is to blame? There is a tendency to seek scapegoats: Blame the teachers or the school board, or teachers of teachers, or parents, or the students themselves, or television. To us, it seems that the whole idea of blame is misguided and unprofitable.

The societal norms and expectations of what mathematics is, of how it should be taught, and of what could be expected from students have evolved through a long historical process. Social institutions, such as our school system, have a tremendous inertial component: Things are generally done the way they used to be done.

That is why a bold look at what possibilities the future may hold if we try very new and different methods may stimulate the creation of educational forms that would not emerge if we only continued to make things a little better by small-scale improvements. Project 2061 has started to formulate some radical new directions that mathematics, science, and technology might explore. Can it get enough public support, in its forthcoming phases, to realize some of those high ambitions?

Adoption of a curriculum, teaching methods, and teacher development aimed at implementing the ideas of this panel report would constitute an educational revolution. How can such a revolution come about? Many elements are needed, of which the most important is a change in the status of the profession of schoolteachers.

Phase 2 of Project 2061 will move from the ideas of mathematics that people should understand to ways of teaching them that will bring them to such understanding. We hereby pass the baton!

APPENDIX

MATHEMATICS PANEL CONSULTANTS

Zvonko Fazarinc Consulting Professor of Electrical Engineering, Stanford University; Director, University Relations, Hewlett-Packard Company

Lyle Fisher Codirector, Bay Area Mathematics Project, University of California School of Education (Berkeley)

William Kahan Professor of Computer Science, University of California, Berkeley

David Logothetti Associate Professor of Mathematics, Santa Clara University

Joel Schneider Content Director, Children's Television Workshop (New York City)

Alan Schoenfeld Chairman, Division of Education in Mathematics, Science and Technology, Graduate School of Education, University of California, Berkeley

Elizabeth Stage Director of Mathematics Education, Lawrence Hall of Science, University of California, Berkeley

NOTICE

This is one of five panel reports that have been prepared as part of the first phase of Project 2061, a long-term, multiphase undertaking of the American Association for the Advancement of Science designed to help reform science, mathematics, and technology education in the United States.

The five panel reports are:

- *Biological and Health Sciences: Report of the Project 2061 Phase I Biological and Health Sciences Panel*, by Mary Clark

- *Mathematics: Report of the Project 2061 Phase I Mathematics Panel*, by David Blackwell and Leon Henkin

- *Physical and Information Sciences and Engineering: Report of the Project 2061 Phase I Physical and Information Sciences and Engineering Panel*, by George Bugliarello

- *Social and Behavioral Sciences: Report of the Project 2061 Phase I Social and Behavioral Sciences Panel*, by Mortimer Appley and Winifred B. Maher

- *Technology: Report of the Project 2061 Phase I Technology Panel*, by James R. Johnson

In addition, there is an overview report, entitled *Science for All Americans*, which has been prepared by the AAAS Project 2061 staff in consultation with the National Council on Science and Technology Education.

For information on ordering all six reports, please contact Project 2061, the American Association for the Advancement of Science, 1333 H Street NW, Washington, D.C. 20005.